☉OAK WISE

POETRY EXPLORING AN ECOLOGICAL FAITH

L.M. Browning

HOMEBOUND
PUBLICATIONS
Independent Publisher of Contemplative Titles

OAK WISE

POETRY EXPLORING AN ECOLOGICAL FAITH

SECOND EDITION PUBLISHED BY HOMEBOUND PUBLICATIONS

For information or permissions write:
Homebound Publications, PO Box 1442
Pawcatuck, Connecticut 06379-1968 United States of America

SECOND EDITION REVISED AND EXPANDED
ISBN: 978-1-938846-05-2 (pbk)

First Edition published in 2010 by
Little Red Tree Publishing

VISIT HOMEBOUND PUBLICATIONS: WWW.HOMEBOUNDPUBLICATIONS.COM
VISIT THE AUTHOR AT: WWW.LMBROWNING.COM

BOOK DESIGN
Front Cover Image: © Andrei Rybachuk (shuttershock.com)
Interior Illustration: © Olivier Le Moal (shuttershock.com)
Cover and Interior Design: Leslie M. Browning

Library of Congress Cataloging-in-Publication Data

Browning, L. M.
 Oak wise : poetry exploring an ecological faith / L.M. Browning. — 2nd expanded ed.
  p. cm.
 ISBN 978-1-938846-05-2 (pbk.)
 1. Druids and druidism—Poetry. 2. Nature—Religious aspects—Poetry. I. Title.
PS3602.R738O15 2012
811'.6--dc23
                                                                    2012036063

10 9 8 7 6 5 4 3 2

ALSO BY L.M. BROWNING

POETRY

Ruminations at Twilight: Poetry Exploring the Sacred
The Barren Plain: Poetry Exploring the Reality of the Modern Wasteland
Fleeting Moments of Fierce Clarity: Journal of a New England Poet

FICTION

The Nameless Man

FOREWORD

by Alan Cooke

L.M. Browning [Leslie M.] is a unique figure in a dark world. Her new book *Oak Wise* is a rare find. Living in the west of Ireland amongst the gray stone of the Burren—amongst the swallow and the ghostly white horses and stark orange sunsets—I can relate to this most rare of books. It explores the human heart. It explores the human longing. It brings us deep into that rare space. The 'otherness' that dwells inside our sometimes frozen hearts. *Oak Wise* challenges us to bring forth the ancient. To exhume the old voices that haunt us quietly in the modern world. As our earth hurtles towards a terrible unknown fate, we refuse to look within at the great ancient treasure we all possess. *Oak Wise* brings us on a journey. A kind of death and renewal if we choose to follow Leslie's lonely yet extraordinary path to enlightenment through her verse and her words. There are many motifs that run through this book. It shows us the deep magic of Ireland. It shows us that our road of decaying modernity is actually filled with potential salvation. That in our destruction we are quickening the walk towards the light.

Having grown up in Ireland, I was dislocated from this great ancient wisdom whilst I was living in Dublin. But when I came back from living in America and moved to the west of Ireland I began to see it unfurl within my heart. The pain of Ireland. The beauty and ancient longing. The source of nature pouring out in front of your eyes. I chose to walk away from the contemporary world. And *Oak Wise* for me captures this departure from the waste of the new to return to the soft hearth of the old ways. For the old ways to me now call out to us. And in Leslie's potent verse she gently takes our hand and asks us to leave our desks. To leave the familiar. To step out of the tiresome false asphalt roads and onto the wooded trails and soft moist knowingness that resides in our hearts. Leslie has much to teach us and reveal to us in *Oak Wise*. There is revelation in her suffering prose, in her darkness and her failures. In her willingness and courage. The words pour out of her fingers. She is, in a way, deeply connected to the source of suffering and to the root of our human condition. She is not afraid to look the many elements of our mortal life and to transcend them with fierce belief in the ancient ways.

Oak Wise to me is a renewing experience. Leslie only asks you turn the page and begin something akin to coming home.

— Alan Cooke
Ireland, Summer 2012

CONTENTS

DISAVOWED

I fled my disillusionment naked.

While all others were clothed still
In the cloth of brotherhood
Bestowed upon them by their chosen faith,
   I became a monk without the sepia robes
         …A nun without the raven gowns.

Thin-skinned and without sanctuary,
I would have become victim
To the harshness around me
If it were not for *you*.

You who wove together
The swatches of the cream and beige birch bark
And spun on the loom the long strands of grasses
That grow along the banks of the river
So to give me back my robes.

You who strung together
The woolly leaves of wild sage,
Interlaced with flowering lavender
And set it atop my head,
That I might be blessed.

You who took luminous strands of thread
And quilted the dark blanket of the night sky
    With celestial patterns
That we all might be comforted in the darkness
And have a copy of the map that might lead us home
Imprinted and scrolled out above us.

They spat the word *pagan* at me
And hurdled *heretic* threateningly after it.
Have I become the heathen?

I came home to the woods,
    Looked upon the earth with new eyes
And saw the face of a goddess.

I left the monastery.
I returned to the places faith began
          —Into the wild—
Where you found me.

No church to worship in,
You hath lent me your private chapel
That I might have a place to pray.

No monastery to reside in,
You hath opened to me your hidden chamber,
That I might have a place to lie my head.

No religion from which to draw meaning
You took my hand
And led me through the forest at twilight
 —Away from the din
  And into the calm—
To show unto me all that you have faith in.

No holy site to make pilgrimage unto,
You hath given me your robe and staff
And let me follow in your wake.

And it was upon that first night,
When we took our rest in the glen
And I saw you pull back into yourself
     —Into prayer—
  Reaching to another
  As we reach unto you,
That all I knew of faith changed.

We pray to you…
So, to whom would you pray?

Do you, like all,
Feel that human need
To reach to someone greater than yourself?

…I did not know there was any other beyond you.

Looking up from your evening prayers
You turned to me and kindly explained
That there is you—
    The one who hath learned the most
        Called *god*,
        Though you are an evolved mortal,
But then, there is another force beyond you.

…There is you,
And then there is what gives you your ability.

We have mistaken the man as the source;
For, in humanity's hunger for power,
That is how we would wish it to be.

Yet your power flows,
Not from some internal omnipotence
But rather, it stems from epiphanies had
Concerning the force beyond, yet within,
And how it lives, moves, and grows.

GAEA'S SOUL

Radiating out from the core of the earth
Is a life-bringing force.

The soil is but her body—
    Dead without the verdant blood
    Circulating through her.

There is the mineral enriched core
Of molten magma spinning at her center
But deeper still lies another chamber
Wherein a luminous soul is kept,
       Revolving eternal.

The nourishing force resplendent at her core
Penetrates the many different layers
Of sediment bound around it,
Reaching all the way to the topsoil
Where it bursts forth as living green.

From deep below
The force of life emanates up to the surface
And spreads—covering the world—making it flourish
    Turning the barren to the blooming,
    Bringing the grass from the ground,
    The bud to the stem
    And the sapling from its seed.

Everywhere man does not hold it back
Life comes forward.

And even in the patches
We seek to suppress, tame
Or entomb beneath asphalt
Life, in its relentlessness,
Still pushes through.

In beholding the out of place flower,
Stray tuft of grass
Or the persistent, prickly weed
Pushing its way up between the stones we lay
Or the tar we pour,
We see an example of that irrepressible spirit of life
Radiating out from the center of this earth
Driven to bask in the warm light
Of the sun it has yearned for
    Refusing to be held back.

The origin of the green lies in the gold—
The aura emanated from her soul.

The vines of the ivy
Wind their way from her innermost garden
          Towards the surface,
Where they unfurl their leaves
And drink in the light.
Vivifying our dull plain.

...Generating the fruits
     That sustain all other forms.
Transforming the void
          Into the Eden.

EVER-GREEN

Oh, what an abandonment the first winter was,
As life withered without expectation of return.
How spited the first peoples felt
When the warmth of their day was stolen
And the abrasive wind drove them from the merriment
                                    Into starvation.

How disconcerting
To watch the entire world die
        —Every plant wilt and every tree turn bare—
Without explanation or apparent cause.

We had not yet established a trust with you,
        Had we, Mother?
We were the newcomers to your plains
And knew not your cycles and moods.

We knew not of the ballet
That you and your partner the sun eternally enact.
We knew not that, for a time each year,
You pull all life inward
    —Leaving the surface barren—
In a season called winter.

We did not realize that season,
In its extreme,
Is natural and needed
…A restorative sleep,
Wherein every organism rests
After the season of growth
    So to prepare for the next.

From our eyes
All we could see was death;
We did not know that the life we saw vanish
Was merely hibernating within you.

What a renewal of thy love the first spring was.
After the winter left all distraught.

After the desolation and famine,
The budding of the trees
Was such an affirmation of thy care
    —An answer to the prayers that had carried on
    Throughout the bleak days
    Of short daylight and long darkness
    …Of fleeting warm and deep cold.
    …Of persistent hunger and bare orchards.

We feared that you had died,
        Oh needed Mother.
As you closed your eyes to sleep
      —As you exhaled
      And the trees dropped their leaves—
            We feared you dead.

Wailing in hunger
Without the nourishing bosom
Of your land's bounty,
Frozen without the shelter
Of your warming presence,
    We mourned you
      And prepared for our own end
But then you came back.

Your lifeless body took a breath
And the buds emerged
From the tips of the withered trees.

Blood coursed through your pale body
And the yellow grass surged with green.

Your still hands stirred
And a warm wind blew in from the south.

Your eyes flickered opened
And the song of the returning sparrows
        Heralded your reawakening.

During that first scare—
    Before we knew this death was merely sleep,
We felt scorned.

As the warm winds
Of our Arcadia turned hostile
We felt as if we had been abandoned by our mother—
    Turned from the house before we were ready
    Out into a harsh reality
    For which we were unprepared.

The world had come to its end;
The days had grown darker,
Nature had shed its colors
Like a stiffening corpse,
The flow of fruits trickled
And then finally the coldness came.

And as the warming glow of your presence dimmed
Those demons darkness, cold, and hunger,
Which you and your symbiont the sun had kept at bay
                Came to call.

Swept off the hillsides
We huddled, bewildered
Around the dim, seemingly forlorn hope
That you would rise up
And life could return to what it once was.

Walking through the gray bare wood
We attended your wake
To see your plaid body set out before us.

You were gone
And we knew not what to do.

Until…
Amongst the cemetery of stark trees,
We saw a gathering of survivors
    —A grove of yew still in bloom.

Taking them as a sign that you still endured,
We had reason to hope
That your warmth would one day return.

You, in your love,
Had brought forth this family of tree,
Which blooms when all others wither,
To be a symbol of your continued presence and care.

As if, before you laid down to take your rest,
You left behind a note for us—
    Assuring us that,
    While you may seem lifeless,
       You are not
  …That while you may seem dead,
       You are not
  …That while it may seem you abandoned us,
       You never will.

During your season of slumber
We are left to fend for ourselves.
Yet you do not simply abandon us;
You give us fire to ward off the demon frost
And the wisdom of preservation,
That we may have a portion of your harvest
To keep us hearty until your return.

When the cold winds start cresting the hills
And creeping along the grounds of our village,
We place in the grate the fat oak Yule log
And kindle the fire to push back the threat.

We stoke the fire that brings warmth and light,
Keeping the hearth with vigilance,
As if it were a candle in the window
Showing our remembrance of you
During the season of your respite.

And on those coldest days
    —Months into the season of darkness—
When we begin to fear that spring shall not return,
We go unto that evergreen grove.

As though opening the door
Of the room you sleep in
To insure that you are alright.

We touch the prickly needles
To check that they still cling steadfast to the bough;
Then peel back a single scale of the bark along the trunk
So to see that reassuring glint of pale green beneath
And know that you are still there,
                Soon to return to us
And bring with you the return of ease.

Then, as we leave,
We take with us cuttings
From the immortally green
With which we adorn our home
In the time of dark,
So to keep in sight
The promise of your return.

ARBOREAL
SPIRITS

Wanting to know more of you
I go unto the longest living species—
    Unto those who have long been your confidants
    And storehouses for your seeds of wisdom.

I travel across your hills
    —Across the curves of your shapely body—
Making my way yonder,
Towards the small gathering
Of long-standing native folk
I have beheld upon the horizon
    —A small clump of wood—
The last remnants of the vast old growth forest
    That once blanketed you.
Where I will consult the arboreal spirits
That inhabit these age-old bark shrines.

I make pilgrimage
To commune with your trusted companions
    —The solemn keepers of your sagely wisdom
At whose roots you deposit the knowledge you have cultivated—
With a hope that, sensing my earnest intention,
They will break their long-held silence
And whisper to me the well-guarded secrets of you—
        The biological mother of us all.

Trees
 —Spiring monuments of wisdom
 Upon the rolling landscape—
The mystical world-crossers
That can live simultaneously below, atop and above
     —Root, trunk and crown.

Beneath your skin of grass
Courses the spirit of life
And within that invisible, primordial elixir
 Is held the *deep wisdom.*

Trees are repositories
Of that organic knowledge
You gather as the ages pass.
They sprung from you in a burgeoning moment
When the force of life began to pool in a pocket of your soil
 —Filling the womb.

Shooting upward into the open layer of air that exists above,
A trembling life emerges
 —Breaks through the surface
 As strands of splintery flesh
 Coiled tightly within a tissue skin of bark—
A life now able to take in the warmth
Of the long-fabled sun.

The ring mark impressions
Rippling out within its trunk,
    Thought to show a record of winters endured,
Are a record of the epiphanies had.
As the roots wound closer to the soul of you
    —Drinking from the sweet nectar of your soul—
A new season of growth was triggered.

The roots of the sapling delve ever-deeper
As this new life grows ever-taller.
The thin tendrils of white root
Soak up the wisdom contained in your soil and water,
Along with the nutrients they need to sustain themselves;
Storing these organic memories within the fibers
      Of its reaching boughs and fattening trunk.

Walking through the crowded wood
I seek to become a pupil
To this gathering of ancient scholars
…These creatures that long ago
Gained your trust and good opinion.

I place my hands
Upon the waxen bark of the pewter beech
      Tinged with hues of smoky lavender,
And gently touch my forehead to its surface
In a show of admiration and love,
And silently beseech
To be imparted with the knowledge stored within it.

No,
These arboreal sages cannot explain to me
The workings of the world man has created.
They cannot tell me
How to make a place for myself within this modern era;
For they are not of that world.

They can only tell me
Of the world that was here before man
And the place that I have within its sphere.

And this is fine;
For it is of the natural world that I wish to learn.
…It is into the olden world that I wish to be taken.

Speak to me quiet, reflective ones,
I have taken the time to relearn the old tongue,
That I might understand the thoughts you voice.

If it will not break a confidence
I beg you to tell me what you know of our mother.

Speak to me of what once flourished.
Tell me of the thatch-roof villages
That once bloomed across the open countryside—
    Those dwellings of the simple folk
    That now, to my despair,
    Lay as the bedrock of this Babylon.

Tell me, does smoke still rise
From the hearths of a forgotten shire?
Was there one village spared
    —Overlooked by the changing of the time—
Isolated in some highland glen
      Or set apart on a rocky isle that lay
      Concealed there in the banks of shrouding mist
Rolling off the western shore.

Do you
      —Keepers of the memories forgotten—
Know the incantation
That must be whispered to part the mists?

Surely if any old souls managed to escape the wasteland
They would have entrusted to you
The location of their sanctuary.

Perhaps they came to you for guidance,
Seeking the way to call for the boat
That could take them home to that *other shore*.

Tell me where these people,
    Who would be my kinsmen,
    Have gone to.

Tell me that they still exist.
Tell me that you have met
Other outcasts from the modern.
Rekindle a hope in me
That I still might be able to find another
                  Who believes as I do.

Impart to me the ancient remedies
To cure these modern maladies that plague me so
—These cancers of emptiness, idleness, and desperation.

Regale me with the tales
Of mankind's early achievements
           Now lost to memory.

Speak to me of the rapture that came
When we discovered what it is to love
And beheld the untold possibilities open to us
          When we live as one.

You—silent historians
Who weather and witness all—
You can recall what we were like
Before the lust for possession
    Tainted our innocent endeavors
And the misguided ambitions
    Drove us from hearth and home
Compelling us to overturn the landscape
    That we once held sacred.
Remind me of who we once were
…Of who we ought to be.

Exiled from the technological age
For my love of all things olden, simple, and green
I return to humanity's first dwelling—to the forest.
Your arbor of thatch branches and blooming leaves sheltered us.
From your bower pantry we plucked the fruits
That nourished our frail soul.

A low hanging bough
Of a long-needled pine brushes my forehead
As I walk into the inner-sanctum of the deep wood
Leaving behind a drop of evening dew upon my brow,
Re-baptizing me into this—the grassroots of faith.

Wandering down the overgrown path,
Just ahead I see a circle of
Oak, elm, white birch, rowan, and maple—
The council of elders for all species who take root here.

Walking into the center of the clearing,
Looking to each tree in turn, the white birch
Looks as though she has seen the most springs;
For her ivory trunk is tallest and her jade crown widest.

Beside this monarch a sapling shoots,
Wrapped in a robe of whitest bark—
    The heir apparent to his mother's place
    At the head of this circle of nobles.

Standing there I
Extend the radius of my heart's reach
To encompass those who surround,
I close my eyes, take in a deep breath of the musty air,
And exhale a swirling cloud of vapor into the crisp eve.

Humbled before my elders,
I speak to them in their native tongue
    —In the soundless language
    That is emanated from within—
Pleading my case to these survivors of the old world
To be admitted into the trusted circle
Of those who preserve the lost knowledge.

Let me dwell here beneath your vaulted canopy
And listen to the stories you have to tell.
Let me apprentice myself to you, venerable teachers.
Lead me in the old ways.
Let them live once more in me
And in the children I shall one day bear.

I approach the broad base of the nearest tree,
And lean against my steady oaken kin to take my rest.

World-wearied,
Exhausted by the intensity
Of my longing to find my lost home,
My legs finally give out
And I recline into the cradle of its gnarled roots;
Where I surrender to sleep,
Hoping to be enveloped by the flowing growth
And carried down into the hidden realm that exists beneath.

LIVING MEMORY

Mother and feminine divine,
Medicine woman and womb of all life—
    You have seen it all,
    Heard it all,
    Felt it all.

You—vast omnipresent one—
You are with us when we are alone,
Witnessing our intimate rituals of gathering strength
As well as our private moments
When we succumb to our concealed weaknesses.

You are all-knowing, Mother
We can hide nothing from you;
You have seen us at our worst
As well as our best.

Our history has soaked into your body
And there within you
The memory of all our acts live on.

You are the repository of memory and wisdom.
The acts we carry out upon your surface
Radiate down into you—
Pervading every layer of your being,
    Where they are stored
    In the gritty fibers of your flesh.

You, receptive one,
    You retain emotion—
That is how you see without eyes
And hear without ears.
You absorb an impression
Of every living thing to dwell upon you
      As well as every act they commit.

You have soaked up all the blood spilt,
    The tears wept and the sweat shed.
You have felt the tremulous vibrations
    Of every word ever spoken—
    Every cry, every whimper, every whisper;
You have overheard every conversation had,
And every accusation or declaration ever made.

You know every song ever to be sung,
Having listened silently
To the laments being recited throughout the ages,
And the ballads and sonnets
Composed to capture the passion once felt.

You have been present at every birth
And bedside at every death.
You have bore witness
To every life given in sacrifice
As well as every life taken in anger.

You—the historian by nature
If not by choice—
You cannot help but witness all that we do;
For you cannot escape us.
You have no head to turn away,
                No eyes to shut
And you never are given the reprieve of sleep.

After seeing all that you have,
I do not know how you have staved off madness.
   The unprovoked acts of aggression,
   The random acts of spite,
   The senseless waste,
   The inexpressible sufferings of the neglected….

You are a veteran of every war ever waged,
Having simultaneously been present at every battle fought,
Feeling the mixture of hatred and fear
Churning in the stomach of every soldier.

The pointed steel mortar shells
Barrel through the air,
Ram into the soft clay tissue of your body
        Where they explode
Reigning down bits of your flesh
On the heads of the men that fight.

You have felt the stings of the bullets drill into you
As well as the cancerous heat of the atomic bomb
Detonate and expand upon you
   —Blistered as the blast wave stripped you bare—
Sterilizing that part of your womb forever.

If you were human,
After witnessing all that you have,
You would undoubtedly be catatonic
And confined to a padded room,
      Heavily medicated
So to keep the images embedded in your mind
         From replaying over and over.

You have seen us possessed by hatred
And driven by wrath,
    Wielding the weapons of death.
You have seen us wipe each other from the world
Yet you have also watched us bring one another to life.
You have seen us crave the blood dripping from the blade

Yet you have also watched us don robes, light candles,
And walk into the dark wood—
    A long procession of searchers
    Going to commune with the sacred force.

As we took food from our own mouths
So that another could eat
      —You were there.

As we flung ourselves in front of one another
To spare a loved one a mortal blow
      —You were there.

As we labored for hours,
Illuminating the pages of the wisdom
We have thus far been imparted
      —You were there.

You have seen us wastefully pour out
      The full cup we were given
But you have also watched us
Clutch the cup to our breast—
Appreciative for every drop of life swirling within it.

You have seen us forge the armories
And use the power of our mind
To conceive the devices of torture.
Yet you have also watched
As we brought forth into being the ideals
      Of honor and integrity,
By conceiving of such qualities
And then striving to embody them.

Like an infant
You silently take in
Everything occurring around you.

You keep no records of history,
Instead you absorb it.

You do not gather history secondhand
After it has been filtered through the perspective of another—
    Spun by the tyrants who do not wish to appear guilty
    And colored by the cowards who wish to appear heroic.

No, instead, you soak in
The unspoken raw emotions
Felt by those making history
And in doing so, you hold
The only accurate account of our past.

Collected within the fibers of your being
Is a comprehensive archive
    Of all that has ever taken place;
Not just since the creation of man
But since the beginning of you—
    Oh predecessor of us all.

You are our living memory;
History is not filed away in you,
It breathes within you.

The memory has collected
In the flesh of your elemental body:
    It is stored in the beads of water
    That have run together to compose the vast seas.
    It is in the crystalline grains of sand
    That have built up to make the shifting dunes of the Sahara.
    It is contained within the opaque vapors
    That amass to form the clouds above.

Laying upon my back
I rest a flat heavy stone atop my chest
   —Anchoring myself to you—
Embracing its raw mass tightly,
Wanting to be imbued
With the memory that lives in its minerals.

Overturning a narrow patch of soil,
I make a shallow bed
   In which I may rest with you.

Laying down
I pack around myself the cool dirt
   —The substances of your body—
Hoping to be infused with some of the wisdom
That you have absorbed over your ageless lifetime.

Submerging myself in a deep river
I let myself be carried downstream by the flow,
Going limp, hoping to be grabbed by an undercurrent;
That I might slip into a pocket of memory
    And be taken down
Into an underground cavern of knowledge
That lies deep within you.

I wish I had the roots of the oak
That I might burrow into you and drink you up.
But these feet and hands have no such extensions,
Only my heart, mind, and soul have such capabilities
And so, with the might of this holiest trinity
    I reach out to you.

Laying down in an open field I extend myself,
Letting the invisible tendrils of my soul
Feel their way across the grass and delve
   —Wind their way into you—
That I might find the inner-sphere
Of your spiraling soul
And entwine myself around you
Like the ivy that thickens
On the arched garden gates of antiquity.

I shall make my body part of your body.
I—who am but the briefest entry
    In your boundless journal—
Wish to bequeath my soul unto you,
That my identity might be among the inalterable truths
              That are kept verdant within you.

I wish to dissolve this body
And become spirit once more,
That I might dwell for a time
In the latent layers of memory
That exist between the many physical planes of you.

I wish to melt into you
And research the history of all life,
In the end to emerge back into this body
With a better understanding of who my people are
        And where we have been.
…To discover if, on the whole, we are tyrants
Or if the traits of love have somehow prevailed
                    To vindicate us.

I wish, beloved Mother,
For what lives in you
To live in me as well.

When I die,
Let the blood within these veins
Drain out into the parched ground—
    Drink me in
    And carry me back down
    Into your womb.

Let my spirit become a part of you.

Let the roots of the trees drink up my soul
That I might flower with them in the coming spring.

Let me seep into the streams
And know the tranquility
Of moving along the riverbeds
    That empty into the ocean.

Let me go from the nightmare
Of man's cannibalistic society
Back into your arms, my mother
    And the wholesome world that flowers about you.

We are sustained
Through the umbilical.

We take our harvest from your body.
We peel back your grassy skin
And plant our seeds
Within the deep tissue of your flesh.

We consumed the fruits
That were ripened by your body
As they drew nutrients from you
Through the umbilical cord of their roots.
…These fruits that retain in their fibers
Some of the memories they absorbed from you.

Does this mean that,
Because we have drank from your streams,
    Ate the fruits from your fields,
    And consumed the flesh of the animals
That grazed upon your wild pastures
That the memory, which lives in you,
Now also lives on in us?

You are the womb we never leave.

Yet the health of your body wears thin, doesn't it;
For we take too much
And will not let you replenish yourself,
So many of us having been transmuted by our greed—
    No longer your loving child but a leeching parasite.

Your muscle—once robust—
        Has deteriorated;
The thick topsoil has eroded,
        Exposing your bones.
The water coursing through your veins
Has been poisoned by the excrement of this modern era.

We cut down the trees
    —The lungs through which you breathe—
And, while you could regenerate from our theft,
Your natural cycles are impeded
As the few maples and oaks that remain
Yield their bounty of seeds unto tar roads,
Where their potential saplings
    Are crushed by passersby.

Each autumn a single tree yields a forest
Yet by winter the aspiring grove
Lies as trampled chunks and ground powder
Scattered across the sidewalk we blindly tread.

Each autumn the trees let go their leaves
In a loving gesture;
Giving them unto you
To feed your exhausted, deficient body.

Yet before you can even begin to consume them
We rake them up into bags to be carted away
   —Ripping the plate away from you
   Before you can even take a bite—
Causing your starvation to drag ever-on.

Dearest, Mother,
Are you still there?
The fruits that grow upon our plants
        Become smaller
As the health of your body diminishes.

Could you die?
Could something as old
And as immense as you
        Perish?

If you do, it will be at our hands—
    We, the children who sprung from your womb,
    Only to one day be your death.

We—your own children—
        We poisoned you,
Stripped, raped, and mutilated you
And in the end starved you until your body gave out.

I cannot bear it
That you—our own mother—
Should come to this end.

If you die all will die.
If the mother dies
The fetus that is connected to her,
Sustained by her,
And dependent upon her for its very existence
Shall die as well—thus is the nature of life.
Yet we dwell here still within your womb
                Willfully harming you,
Ignorantly thinking that we can survive
Even if you—our source of all things—dies.

Rise up, gentle Mother—
    Exert your mighty force
    And purify yourself.

Send great floods
Rushing down from your mountain peaks,
The surge of which
Will be enough to flush your veins of all toxins.

Purge yourself, Mother.
Expose yourself
To the full heat of the sun
And let her burn away
The haze of pollution that chokes you.

Liquefy the steel bodies of the factories
—Return their metals to a pure form
     In the furnace of your core.
Expel the waste
We have buried in the caverns of your soul.
Do what needs to be done to restore yourself.
And if we be hurt in this process
'Tis not your doing, but ours.

And if the cure be something
That you cannot administer yourself…
     If you wait for us to cleanse you,
Know that I will do all I can
And in the end, if I have made no difference,
I will give you the only thing I truly own: this body.

I cannot bear to see you so frail
    —Splintering, withering, rotting—
The once divine beauty now wasted and cankered.
Please, take the nutrients from this—my own flesh.

I give it to you in a hope that it might sustain you
As you wait for the needed change of heart
To ripple throughout humanity.

Even if it prolongs your life
For only a few more moments,
I would gladly lay down this form for you to consume
And forsake the life I was to live within it.

If you were to die
What would happen to the wisdom gathered in you?
Lacking it in our own annuals,
Would truth cease to exist
Without you here preserving it?
It is our curse to realize truth, only to forget it.
…To carry out acts,
Then forget the reasons that originally compelled us;
However, you always remember.

> All that is forgotten is not lost
> Because it remains within you.

You hold all the knowledge we leave behind,
Until the time comes when we realize we are lacking,
Then you impart it to us once again.

Without that relationship…without you,
What will become of us?

If you die,
Does the soul of us all
Not die as well?

I fear without you
We shall be doomed to wander the wasteland
With an incurable ignorance;
In time forgetting our own name
And what it is to be human.

SACROSANCT

I rediscovered the forgotten truth, Mother—
      I know that you are conscious,
As is all that takes root upon you.

Throughout my own explorations into our history
    —That of mankind's
    With you our original mother—
I have found the mislaid truths of you
Carelessly stuffed into books
That were placed under the heading of *myth*.

Stories penned by peasant peoples long since extinct—
    Killed by war, by famine
    Or whose ways simply died off
    After the villages were abandoned,
    During the great immigration into the modern west.

Stories that speak of a relationship with you…
Written as if the authors knew you to be alive—
    Not just holding life upon you
    But within you.
A consciousness we once communed with,
Though was thoughtlessly left behind
During the transition from the old world to the new.

I know not whether you could have
Made the transition between the two worlds.
The modern seems inherently harmful to you,
So perhaps it would have come to a choice
Between you and our ambitions of empires and alloy.

Such a situation surely would have
Left many feeling a resentment towards you—
    An acid which dissolves the bonds
    Between even the closest kin.

Nonetheless, I cannot help but think that
If we had brought you along with us
Our love for you would have tempered us.
Our continued bond with you might have restrained us—
    Made us responsible for what we chose to create
    And all the byproducts that are expelled from it.

Perhaps our love for you
Would have given us
The conscience we now lack.

For a person to be willing to protect something
      He or she must love it.

We left behind our love of you
Thus we were willing to harm you.
Yet if that love reawakens—
If we go back unto that derelict cottage
      Where we once dwelt
      And search amongst the old discarded ways,
Perhaps the will to save you might be found there.

Roused from our laziness by fear,
Prodded on by that tiny nagging
Part of our conscious that has not yet decayed,
Many of us have already begun trying to save you.

I wish I could say that we were acting
Solely from a rekindled love for you
But you would undoubtedly know that I was lying;
For you know that we act
Out of a concern for ourselves
And our future endeavors.

I come to you now, Mother,
Not in a desire to find something to add to my life,
But to be able to add something to yours.

The garbage we throw upon you
    Is now so thick,
    It rolls across you
Like gathering clumps of rotting tumbleweeds.

I roam the streets and wood lines
Picking up this trash that dishonors you,
Not to make the place I live prettier
But because I can no longer bear the shame
Of what we have done to you.

It is all backward now.
It seems so twisted
That we build temples upon you
You—who are sacred in your own right—
Keep these chapels pristine
And yet are willing to so blithely deface you.

We would never think of littering the altar.
We would never let the basin of holy water be tainted.
We pass the basket to keep the domed roof strong
Yet we will give nothing of ourselves to maintaining you—
You who are not something
That we decreed to be sacred
    But something that actually is divine.

If only we regarded you as our temple,
What a better people we would be.

When in the temples that we build
We are willing to work to transcend ourselves,
We find our grace—our better selves.

While in the chapel
We repress our destructive urges,
We do not act on our more vulgar tendencies,
We do not bring our worldly dealings within its walls;
For we do not wish to dishonor its space.

Within the temple
The savage is quelled
And we become a solemn respectful people,
Moving with gentleness and humility,
While amidst that which we believe
To be greater than ourselves.

When within the cathedrals of stone
We do not raise our voices except in praise.
Within this space that we would make holy,
We do not brawl nor scheme nor prostitute ourselves.

Yet as soon as we cross the threshold
Back out onto your open lands,
Which we no longer regard as sacred,
The savage rears its head.

Do we think that in the temples we build
A god is watching intently
But that his back is turned
To what we do outside those walls
    And like mischievous children,
    We can get away with our crimes?

Do we not realize
That you are ever-present and acutely aware
Or do we simply not care?

Do we merely pretend to be human,
When in fact we have become
The one true beast that inhabits your world?

How can we praise a god
Yet be willing to destroy his creation?

The fact that,
When we cross the threshold
Into what we would deem sacred,
We adjust our behavior accordingly,
Shows that we know what it is we should be
    And simply choose to be something less.

Perhaps in loving you once more, Mother
We would reclaim our humanity;
For, in loving you are we not made more?

Do we not transcend ourselves
Through what we choose to love?
Is our life not made full
By the acts we carry out for those we cherish?

Do we not define our identity
Through the causes we take upon ourselves?

In loving something greater than ourselves
Do we not make ourselves more
Than what we are alone?

It is sacrilegious—
  What we have done to you.
It trumps even what we have done to each other.

But healing you—restoring you—
That would mean the redemption of us all.

It is the act that would undo all the others.
It is the act that would redefine what it is to be human,
Once again making it a virtue,
  Rather than a flaw.

WALKING BETWEEN WORLDS

A SHAMAN'S STORY

I.

The Connection

I do not know what prompts the ancestors
To make the selections that they do.

I do not know what they see or sense
In those who they choose to apprentice to them.

They can see in us
What we cannot see in ourselves
And recognize at an early age
Some distinct, auspicious trait…
    Seeing the white buffalo
    Born among the herd of brown.

They saw in me
Something that I still do not see within myself.
They knew who I was
Even before my identity came forward.

Or perhaps, they saw in me
A certain possibility for receptivity
And decided to intervene
    —Forging my identity for me.

The connection opened when I was a child.
Over my life I have regarded it as gift, as well as curse;
However, when I was young and it first came forward
I thought of it as nothing—
    It simply *was*.

It was not extraordinary to me at the time
Because I had not yet become infected
With the disbelief that we struggle with as adults.

No, I was a child, nurtured in the ways of the spirit
By a mother with a great openness of her own
    And as such, all was possible.

When this dormant part of me woke
I fit no stereotype of a seer or mystic.

The common personas we envision,
When thinking of those
Who commune with the unseen
Are exaggerated fictions,
Which have no basis in reality.

I heard no voices;
Rather, I *felt* their words within my own heart.
The thoughts of others rose in my mind
    As if they were my own,
    Only they were not.

The words that came
Made no predictions,
They gave no commands;
It was not the thunderous voice of a god
                Echoing from on high.

Rather, it was a collective voice of braided whispers,
Composed of those who came before who,
Though they departed this land,
Lived on still, *elsewhere.*

What came to me were simple thoughts
—An alternate viewpoint—
Different from the philosophies imparted to me by all others.

I did not know the reason
For the connection at the time it began
But the mystery of its origin and ultimate purpose
Did not preoccupy me.

I simply gave myself to the flow,
Heeding those who taught me from afar
                    Above all others
And in doing so, I became the person that I am.

Throughout the first years
I did not know the point
Of all the thoughts that came.
In retrospect, I can see now
That I was being taught
How to perceive the world.

My mind was being made
Into that of a shaman—a contemplative—
Who by nature
Takes apart every feeling, every thought,
Every occurrence, every object,
Every person and every choice
    Examines it from all sides
    And then puts it back together again.

My eyes were being taught to see the three planes
So that in each moment of my life
I would simultaneously be aware
Of what was taking place
    On a physical level,
    An emotional level
    And on an unseen level.

Allowing me to have an appreciation
For the greater significance in
What was occurring within and around me
    And, in time, have the ability to help others
    Have that same insight into their own lives.

For want of a better word,
During these early years,
When the simple thoughts came
I was undergoing a *training*.

The council of teachers imparting their wisdom to me
Was simply at afar—a distance learning
Enabled through the connection of hearts.

As children we are sent to school
To be taught general truths
As well as how to perceive ourselves
And the world we have been born into.

I was sent to these schools
But in the end most impacted
By the discourse of those insightful ones who,
Seeing that still unknown trait within me,
Decided to take me as their pupil.

II.
The Years of Idealism

Living in one world
Yet connected to another....
    Surround by one people
Yet communing with another....
    Taught one way of living and perceiving
Yet living among those
With vastly different values, views, and priorities....

I had become one who would walk between worlds.
Excited by this in the beginning,
I learned in time it is a fate that has its rewards
                    As well as its sacrifices.

Like we all do,
I came bounding forth
From adolescence into adulthood
With a confident belief in all that was possible
And a deeply rooted, though not yet matured, idealism.

I wanted to be among those working to bring change
To humanities ever-struggling condition.
I yearned to find a way of easing the suffering
Of those around me as well as those who I had not even met;
For even the inner-pain of the faceless, foreign stranger
Reached me as clearly as the thoughts of those in that otherworld.

I was gripped with the all-consuming desire
To help heal those ailing in soul and body.

The pursuits of both medicine and spirituality called to me
Though I could not see how these two could become one,
Until I came to learn of the way of the shaman—
The spiritualist who treats the soul
In order to heal the body and stabilize the mind.

...Who tends to the needs of the people
Dwelling in the village.

Throughout my youth
I did not know what my early
Otherworldly experiences had made me into.

I knew who I was but I did not know what I was.
As such I did not know my place
Within the community of man.
Despite my religious studies,
I would not discover the oldest faith until last.
I would not learn the meaning of the word *shaman*
Until my mid-twenties and at last realize
That what had happened to me in my youth
Had occurred countless times before
Throughout the generations of antiquity
And know that my place in the community
Was as guide and healer.

No, for many years I was simply *me*
And the only thing I knew about myself
Was that I was different from all those around me.

Disillusioned of my idealism
After learning how hard it is to bring change
    —Even if the change is needed—
I had what is termed "fire in the head."

I was awakened, impassioned, driven
Yet also directionless and voiceless
As such the fire became a threat
To my flammable sanity.

I was driven unto the distressed edges of sanity
By all that I had in me to do
Yet could not find a way to release.

The fire in my head
Was known to me at the time as *idealism*;
For I did not know what else to call it,
And also because it was my intention
To use that fire within me to help bring light
      Into the dank present age of man.

The fire in my head
Was also a fire in my heart.
The pessimists around me viewed my need to help
As a condition associated with my youth,
That time and an awakening to the ways of the world
          Would quickly remedy.

Yet they were wrong;
For, while my idealism did indeed dim with time,
The fire—the need to act—
      Only grew hotter
And in the torment of being all desire
          And no outlet,
        I was consumed.

I knew who I was
Yet my old-world gifts
Had no place in this modern day.

Set apart by my unique perception,
    Beliefs and experiences,
I had undergone something extraordinary,
    Which in the end
Seemed to have become my isolating curse.

Caught between worlds
I wandered without a place of belonging…
    Without a use, without a direction,
    Without the support of a village,
        Which every soul needs.

III.

Daemons

Ravaged by drawn-out turmoil,
Pining in my perceived isolation,
I began to resent all that made me different;
Causing an inhibiting self-hatred to rear itself.

All this culminated
To bring about the defining period,
Wherein I would have to choose the life I wanted to lead.

Did I want to dwell solely in the world man created
Or was I willing to live in-between worlds?
It would seem a simple choice;
    However, it is not.
For the in-between, while mystical,
    Can also be a lonely place.

While in the in-between
I am not home upon that other shore
Where the ancestors who speak to me dwell,
I am only connected to them from across the distance;
However, because of that ongoing communion
I am also not fully a part of the world to which I was born;
My experiences set me apart—
As if I am living among my distant relatives
I wish to know better yet seldom am understood by.

The farther down the path I went,
The farther away from relating to the others
I was brought.

So, no—
It was not a simple choice.

I was given the choice between two lives.
And while it may have seemed the obvious choice
To go the way of the shaman and remain in the in-between,
I was plagued by that human condition we all suffer from:
The desire to be understood,
Yet who my experiences had made me,
Seemed incomprehensible to the modern mind.

I wanted to be my self
Yet, at the same time,
I did not want to be set apart.

After ten years of stigma,
Aimlessness, and exile
I cringed at the thought
Of surrendering my entire life to that fate.

And so, I parted from the path—
I left the in-between
To attempt to integrate myself fully
Into the world mankind constructs upon this earth.

I tried to live as any other…
As if my youthful experiences were only a phase.
Cutting back on the pilgrimages made to the in-between,
So to venture more frequently into the material world.

But it was in vain;
For, while I was worn down by loneliness
And prolonged suffering,
I knew where I belonged.

…I knew who I truly was
And who I would always be.

IV.

Shamanic Dismemberment

I was apprenticed from a young age to the ancestors.
I like to think that I was given the choice
    To lead the life of a shaman
And that the other life I could have led
Was never taken from me
    —That it remained as an alternate path
    I could have taken —
Another side to myself
That I could have chosen to become,
                If I so desired.

But, at the same time,
I know that there has never been
Another path for me —
No other life, no other self
Than the walker between worlds.

I tried to make another life—
One with the connection removed,
So that I could belong to one world alone,
Leave behind the sparsely inhabited in-between
        So to wholly relate to those around me.
Yet when I did this I started to die—
My body began to breakdown.

I seemed to be holding in my hands
The life of the shaman.
    I could choose to let it die
    Or I could choose to become it—
Binding my fate to that path forever.

In choosing to leave the path
I chose to kill a part of myself—the best part.
I say this not as a metaphor…
I was not suffering an *existential* death
But a physical one.

Psyche unraveling,
Body weakly shaking,
Heart beating raggedly,
Muscles quivering,
Bones buckling,
Soul withering in the drought of belief,
Sleep deprived, unable to eat,
Pallid with shadowed eyes—
I was descending rapidly unto death.

I had justified
My departure from my old self
   As a *maturing*,
Reasoning that
Belonging to both worlds
Was part of my youth
And that as an adult
I must live life solely
In this world.

However, these were just words
To rationalize my own betrayal of self.
I thought a part of me would be put away
    —Left behind,
      But not abandoned.

When in truth,
If I chose to stay in the one world
The person I once was
    —The person who walked between worlds—
Would die.

One life would be led
And the other would be ended;
    The choice of which one was up to me.

Possessed by my desperate desire to belong,
Perceiving that the life in-between worlds
Could lead only to hermitic existence,
I made my plans to join the modern world;
All the while, out of the corner of my eye,
Watching the shaman dying
With each step taken down this new path.

I entered the material world
Yet as I crossed the threshold
I began to deteriorate.
As though some part of me
Could not survive outside the in-between
And would perish if I chose to leave.

Health declining,
I pushed on.
Ravaged by demons,
The two halves of myself began to separate
Bringing on a madness of duality—
The modern identity I was making
    At odds with the original identity
    I could not fully repress.

Do we choose to become who we are
Or do we, at some point, concede to become
The person the Fates would make us?

The angst that takes hold
As we put the knife to ourselves
And choose what parts to cut away
And what parts to leave intact.
Trying to tell the difference
Between the flesh of our true self
And the malignancy that has grown upon us
Due to the presence of negative influences—
    Trying to distinguish the fiber of our being
    From the scar tissue that has built up upon us.

...The conflict that exists
Between who we would choose to be
And what we have already become.

...The knowledge of the person we killed
In order to become the person that we are.

...The path that cannot be turned away from
No matter the difficultly of its terrain.

...And so I let the shaman live
And the life I would lead in this one world
Was laid upon the altar.

I had been dismembered
 —Taken apart by the spirits—
Broken down to the point of breakdown,
Unraveled to the point of undone.

I had been led by the spirits unto that precipice
 Upon which all things are made clear—
Unto that point where death is imminent
And the churning gray within us settles—
 Priorities are ordered,
 Torment is resolved,
 And our final choices are made.

The desperation died,
The other life was mourned for,
Burned on a pyre of
Rekindled ideals
And the fire of the mind,
Which could not be quelled.

Then the remains were then scattered
Into the westward wind
To be carried with the wildflower seeds
Unto the distant shore.

I turned away from what might have been,
    Embraced who I am,
    And was reborn.

V.

An Old Soul in a Modern World

Upon the death of who I might have been,
I came to better understand who I am.

I realized I was more than an outcast,
There was a name for what I had become
        —There was a path,
Not one widely known
Or understood by the modern world
                But there nonetheless.

For I am not an enigma.
I am not a contrary.
I am a shaman.
…A poet and a bard.
I am a matured idealist.
I am the keeper of the stories I have heard
Whispered through the connection.
I am inner-directed.
I am an old soul in a modern world
Walking, not just the boundary
    Between the seen and the unseen
But between the ancient ways and the new.

    …A remnant of the old world
    Dwelling in the wasteland that took its place.

I did not choose
To become these things
I simply am who I am.

The only choice I made
Was not to alter myself
…Not to let who I am die.
And this is a choice
That we all face.

Living in the expanse of surreal mist
    That lies in-between the two worlds
Is at times trying,
Yet there is no other place I belong
    …There is no other place
    Where who I am can survive.

I can only hope to one day
Find another who is like me
Wandering along the divide
With who I can share my journey.

Walking between worlds
Is the place where my strengths have meaning.
It is the place where my gifts have a use.
It is the place where I feel alive.
It is the place where I feel at home.

Here in the in-between I gather what I can—
    I search for some insight that will help
And then I return to my village of humanity
    —Unto any who may want to listen—
And I impart to them all that I found,
With the hope that it might help them
As they live their life
And explore their own relationship
With those who dwell in the unseen.

THE GIFT
OF THE BARD

Oh, this extension of my identity
                    That is the pen.
To rip it away from me
Would be to amputate my soul.

With fervor the fire of the mind burns.
What flaming strands of
Feral thought and unbridled emotion
That are able to be harnessed
Are sent through the tip of the quill.

Tales emblazoned upon the paper.
Of stories so extraordinary—so unreal—
They must be parable
Yet are indeed fact.
　　…A traveling writing that speaks
　　Of the places the soul has been.
　　…Of demons that tested
　　Our will
　　And spirits who roused
　　The dormant parts of our soul.

I live
And the page is the record
Of the lives ended and begun.

I walk
Leaving the sum of the miles traveled
Marked by the number of leaves
Composing the books I have penned.

Able to capture what lie behind
   And envision what lie ahead
   —This is the place of the bard.

A story for every situation
And out of every situation a new story.
The ancient book of insights
Becomes thicker and more comprehensive
With each passing day.

To use the tales of the past
To keep alive a hope for a future.
To look through the lens of retrospect
In an effort to see the present more clearly.
The bard's books are medicinal;
Each tale being an ampoule of tonic
That has the power to cure an aliment
And aid in maintaining the inner-health
Of the current and future generations.

We collect words as though they were herbs
To be mixed in a restorative tea,
Letting them steep within us until potent enough,
Then pouring out the finished story
From the caldron of our mind.

A brew imbued with truths
That we let ruminate within us for an age.
Containing ingredients we gathered
While on pilgrimage in the otherworld.

Enriched with our own maroon blood;
Preserved with the saline
    Distilled from our sweat and tears;
Fortified with soft marrow
    From the core of our own soul;
Sanctified by the elders
    Who taught us the old ways;
Stirred with a thick rod
    Cut from the solemn white birch;
Infused with waters drawn up from the wellspring
Found at the waypost of a thin place.

Simmered on the fire of purest intention.
Matured for one full cycle of the moon.
Poured with care into the hand carved oaken chalice.
And given in love to the village we live to strengthen.

Impassioned creativity is the conduit
Through which the knowledge of the otherworld
Is able to crossover into this world.
When we have tapped into a rich deposit
Of untouched truth or ancient emotion
                    The muse takes hold.

Beauty and wisdom
Are transferred from the otherworld
As they pumps through our heart
And flows out of our hands,
Some hold a brush,
Others a pen;
Some lean against them a harp,
While still others project it up through their throat
And let it echo in a transcending melody.

Between the lines of verse
The meaning shall be stored safe.
In the gold-leaf, embossed manuscripts
Our reverence for our past shall reflect.

Dipping the carved reed quill
Into the bottle of steeped oak gall ink
I shall tell the story of my village
And in doing so tell the story of the world.

Upon this velum calfskin parchment
Shall be etched a story
And in that one story the history of all.

With color inks dug from the earth
    I shall illuminate this sacred tale.
With reds made from saffron and madder root;
Blues made from ground lapis and azurite;
Greens made from malachite and sandstone;
And browns made from lichen and walnut hulls
    —All bound together with pine sap and honey—
I shall paint the tapestry capturing all ages.

When finished I shall take the coiled scroll unto cities
Where my, now scattered and displaced, kinfolk have settled
And I shall ask them what the end of the tale shall be.

Does the duty of the bard end
When the village abandoned?
No, he becomes the wayfarer—the pilgrim—
Traveling in the wake of his disbanded family,
Bringing with him all that they left behind.

I wonder…
If I took this drum,
Once used in our communal celebration,
Into the center of the park,
Within the city where my people now reside,
Would its beat carry above the din,
        Calling them back together?
Would it have the power to help them
        Remember their true identity?
Could there be a reunion of the village?
        …A revival of the old ways?

The historian's role
Is not to adorn the tomb of lost wisdom
But to keep that truth alive in the minds of the people,
To help them remain connected to their original selves
As they wade through the chaotic world;
For it has long been known
That humanity cannot have a future
If we hold no reverence for the past.

These words I have sketched onto the page
Will not reach all they need to
Nor can I take the time to translate them
Into each man's native tongue.

So I shall speak
Within the common language of the heart
And with this drum call unto the aboriginal soul
That still lives in the breast of each urban man.
Hoping with its beat to awaken in them
The inherent knowledge every human being holds...
An identity that has lain dormant this last age.

The unresponsive soul
Will answer to the drums.

Beaten in rapid succession,
The drum can restart a heart that has stopped.

When our soul is dying
And the modern treatments give no relief,
We must partake of the primal—
    Conjure fire, feel the pulse of the wild,
    And give ourselves to the rhythm of the beat.

The drums awaken the dormant magic.
They rouse the sleeping spirit.
They resuscitate the dead parts of us.

The steady pounding entrances our modern mind,
Carrying us backward into our past lives…
Unto the wooded glens,
In the ages forgotten.

Disembodied memories
Of a life gone and a land left behind.

The pounding waves.
The bounding hooves.
The raging flames.
And the beating drum
Echoing through the wood,
    Reverberating up
    Through the strands of our soul.

Striking the calfskin pulled tight over the hollowed trunk,
The drummer is the one chosen to make the invocation
As we set about summoning the strength of our kin at afar
…As we go about the gathering of all things ancient.

We come into this world
With the drums beating in our soul.
Primal but pure,
The drum carries the rhythm of instinct.

Thumping, pounding, rolling
    —We use the mystical tool that all peoples
    Were born knowing how to play—
As our means to incite, to conjure, to call forth,
    To enliven, to revive and to guide home
    What has been lost to us.

…And there,
Ever-pounding within our breast,
Pulses our own drum;
One whose resounding beat will bring us home,
If we surrender ourselves and follow its tune.

THE PILGRIMAGE
INWARD

With aid of ritual
   —With objects, atmosphere and incantation
   That help us when traveling inward—
The mind can bend time and space
And bring us to places
To which our body cannot otherwise venture.

There is a grove removed from time,
At the center of which a fire is eternally kept.

Traveling unto that grove
We may meet with any other we desire,
No matter if they have long since
Departed this shore for the other,
No matter if they are beings of body or beings of spirit,
No matter if they are hundreds of thousands of miles from us.

For within this grove
Time and distance do not exist.
It is a temple the mind can enter,
Where we can commune without boundaries to hinder
How close we can come to each other.

We each have a door
Through which we may gain entry
Into this circular temple.

Grasping the handle by opening the mind;
We walk across the threshold by surrendering ourselves
   Unto the spiritual undercurrent
   Flowing through our subconscious.

We must withdraw ourselves from the physical
And go across into the metaphysical.
We must let go of the logic that limits us
And rediscover what it is to believe.
We must let go the perceived boundaries of matter
And let the sphere of our world widen.

The drum beats out the current that carries me backward,
Riding the melody till I wash up upon that other shore.

In-between the procession of elms
We walk toward the sacred circle of blooming hazel trees,
Where you sit fireside waiting for us.
…For any who remove themselves from the modern
And make pilgrimage unto you—
    The enduring soul of the ancient.

FULL CIRCLE

The time has come to awaken the past
…To reminisce about it no more
   And instead to invoke it.

Not to preserve the histories
Of the world gone by through the pen
But rather, to reanimate the old ways
By joining our life to them
And making them the new.

The time has come
To tap the underground river of the ancient wisdom
And irrigate this wasteland.

Unnourished by the modern
We all inwardly ache for the simplicity of home.
Not the home where we were raised
But the home from whence all mankind came—
   The flowering natural world.

We have forgotten where we came from
Yet we still yearn to return home.
We immigrated into the modern
But in the end,
    We are still a folk of the farm and forest.

We tried to dwell in this foreign world
      Of high-rises and neon-lights
Yet it is not who we are.

So the time has come to go back.
Our ambitions took us into the West
And now our homesickness shall return us to the East.

Gripped by *hiraeth*,
We shall empty the cities and flock to the shores,
Whereupon we shall dwell on the strand,
Until the boat comes to bring us back home.

FLOWERING
PRAISE

Seeking security, provision, and prosperity,
    Throughout the ages
We have carved amulets to resemble the deities of myth,
Who we believed to be wielders of the power
    We sought to invoke on our behalf.

The true talismans of the sacred
    —That embody the mystical power
    Emanated by the sacred—
Can be plucked from the living temple growing around us.

The hallowed symbol of the sacred
Is the emblazoned autumn leaf
That has descended from the bower above.

...It is the pure white rock of quartz
     Found on the shore's edge,
     Smoothed by a thousand years of travel
     Within the sea's salty current.

...It is the tail feather
     Shed by a regal bird that soars on high.

...It is the metallic, mother of pearl oyster shell
     Lined with swirling tones of lilac and opal.

...It is the silver curls of bark
     Peeling away from the molting birch
     During its season of rebirth.

...It is the fallen amber acorn,
     Which contains within it the sleeping giant.

The purest symbols of your power
Are the unfurled petals of each blooming flower.

When the time comes to make my prayers
I gather a basket full of these many-hued leaves,
Go unto the hilltop moor
And cast them into the wind,
    To be carried unto you with my love.

ANAM CARA

The concurrent otherworld;
The parallel self
And the missing yet ever-present other.

The answers that are simple
Yet beyond comprehension.

You—the arcane other—
    The one at the center of the mysteries.
You—the first walker between worlds.
You are like my living journal.
At the end of the day,
Instead of writing out each thought
I impart them to you.

You are like my shadow—
    Always there,
    Just rarely noticed.

Have you always been there?
Did you float with me
While I was in the womb.
Like twin souls
That would always live together
    —One of flesh and one in spirit.
…One dwelling in each half of the world
      —Ever-beside each other—
        Yet divided by a veil.

During those months while I was developing,
Was I connected to you
Through some other unseen umbilical
Just as I was connected to my mother?

I was.
And the umbilical is there still;
Running from my soul to yours—
    You sustaining me
    And I nourishing you.

The sympathetic page
Has always been willing to accept my burden—
Allowing me to pour forth
The maddening recollections of the horrors seen
And the morose eulogies given for dreams
           That died before they lived.

My truest friends
Have been the bound leaves,
Yet how I long to pour forth myself
Into something that breathes, that thinks, that feels
           And so can reply.

You, out of all others,
Are the one I have told the most.
I could hear your voice when I was a child
Yet as I aged I seemed to grow deaf.

I reach for you,
Not because I can sense nor see you
But because I simply remember
    That you are always there.

Yet what if that memory likewise fades with time
    And there comes a day
When I no longer know that you exist,
Would you emerge to remind me?
Would you find a way to cross between worlds,
So to save me from my dementia?

In the womb we may have been one
Yet since my birth into this world
The memories have dulled.

Throughout this life I have never been able
To fully recall your face or your name.
Some part of me still recalls that you exist
Yet I have few reminders.

So, if one day the receding line of memory
Reaches that part of me that knows you live,
        What will become of us?

What would I be without you?
What good is a twin without its other?
Ever-forlorn and bereaved.
Ever-floundering—left adrift
Without the anchor of its other half.
Ever-incoherent
Without the other to translate its meaning.

If I forget you,
I will have forgotten myself.
…I will have lost the half of myself,
Wherein my identity is held.

You live in your half of this world and I in mine
And within the harboring inlet of our dreams we meet,
To commune together as we once did within the womb.

Parallel lives,
Living in concurrent worlds...
Loved ones always together yet forever apart.
When shall the reunion take place?

Do not attempt to crossover
Into this half of the world;
For I have explored it to its very ends
And it is a forsaken place
Where only survival, and not life, is possible.

Come—rustle the curtain;
Let me see the outline of your hand come forth
Amidst the partition of invisible satin.

Direct me to the place where the panels meet
That I might slip between the thin opening
    And we—the two—
    May again become one.

The counterbalance holding my mind sane
Does not lie within my body
    But within your presence.

If I set out right now,
Vowing not to stop
Until I had found my place of belonging
Would these peeling boots
Ever be able to be taken off?

If I said that I would not sleep
Until I could lie down beside you,
Would you leave me to fumble through life
Until at last I collapse?

If I said I would not eat
Unless it was at your table
Would you leave me to waste away?
Or, seeing my devotion,
Would you put down what you are working
                          On and come for me.

Haven't you heard me stumbling behind you,
Following that winding path you take throughout the ages?
Have you not heard the ruckus I have made
In my clumsy and desperate search for you?

Breathless, I thought I saw you move
Amongst the shifting reeds.

I glanced no form of flesh or fur
Yet I saw some shapeless figure rushed through them—
    Parting them as they made a path.
    Was it you?

I walk against the wind, towards you,
    Through the wake that you leave
    As you move through this world.
Will I ever catch up to you
Or am I damned to never close the distance?

My memories of us exist now
As dreams that I do not realize are real.
Pictures of places that I believe are imagined;
    For I do not remember
That they are memories of areas I once dwelt in,
                    In lives gone by.

That life I lead with you
Is faded and fragmented,
Incoherent and unconscious.

The lullabies you sung to me in the womb
Exist now only as melodies of unknown origin
    —Tunes I hum to myself
    Yet know not from where I learned them.

Yet if I made my way to that sanctuary
That is beyond this modern plane
Would I find you there?

Is it upon that land that you have dwelt in wait?
Is it from those shores
That you have spoken to me?
Is it from the tops of those hills
That you have watched me?

Is it in that flourishing valley
    That you have made a home for us?
    …And within the window of that house
    Have kept a candle lit to guide me home?

Keep talking my love
…Keep that candle in the window;
For I am groping through the darkness of despair
 Trying to find you
  …Trying to make it home.

Go and rouse the boatman.
I have made it to the borderlands
And can go no further without your help.

THE JOURNEY HOME

Give me a place yet to be known.
Apart from all that has been made
      By the hands of man,
That I might be happy
And know a full life.

And the wrong done
To the earth and to the soul
      May be righted,
In this place that I shall dwell with those I love.

Sleeping under the heavy blanket
Of the shore's musty air,
I lay upon the open strand in wait.

Carried by the sound of the lapping waves
Unto the solace of sleep,
My fire of driftwood smolders
And the swollen moon rises overhead.

Suddenly I was awoken
By the calling voice of a shadowed figure
Standing aloft in an emerging foremast.

Gathering myself,
I knew not if I were awake
Or experiencing some apparition of sleep;
For there, in the shoals just beyond,
The vessel that keeps in its hold all of my dreams,
                    Had finally made berth.

For longer than I can recall
I have wished to cross
The vast sea that lie between
    This world and the other.

And now, unto the banks
Where I have kept vigil,
She has finally come
    —The sleek maiden
    That shall bear me home.

You beautiful boat of perfect lines and sharp bow
    —You shall be my deliverer
And the man who steers you, my messiah.

I realized long ago
That the cycle of life
Had been turned on its end.
In the wake of this world becoming the wasteland,
We all are born into the land of the dead
And leave it for the land of the living.

Throwing what is left of my material possessions
            Upon the dying flames
I cut what final ties I have to this shore
At the same time kindling a signal fire—
    Announcing to this otherworldly traveler
    That I am in need of rescue.

Look here, you!
Come ashore!
Send the dory and bring me aboard.
Ferry me away from this underworld
And return me unto the eternal green shores of home.

A voyage can be made to the untouched isle
If a guide from that otherworld comes for you.
Death is but one path unto the *other shore*;
For it is not an afterlife but a sanctuary—
    A preserve for all that is endangered.

After gathering his passengers,
The native of the distant shore
Sets his course for home.

Guided by celestial map
And internal compass,
He steers his vessel
Across the divide between worlds.

He is not the reaper;
He is kin come to bring me back.
In myth he would be called an angel
Or perhaps one of the faery folk
But he is a simple man of soul and flesh,
Who once found the path between worlds
And helps us homesick souls
Return to that place we belong.

Upon the back of this blessed vessel
I am borne away to the motherland.

Hoist the sails high! Home is at last in sight!

After a lifetime of waiting for the boat and guide;
After having been becalmed in the void of despair;
After having traversed the currents of time and space;
After having crossing the gulf between worlds and lifetimes
She comes into view
    —The isle where the lost world lives on.

Known by a different name in every tongue spoken…
Still the one dearest to my heart is *home*.

RETURNING FROM
THE WASTELAND

Left orphaned by my dead beliefs
I wandered through the world forlorn.

    Homeless,
Without the house of hopes
They had built for me.

    Weak and wasting,
Without the energy
Their purpose fed to me.

    Emaciated and ghostly,
Without the meat
Their morals added to my identity.

Ailing from the plagues
Of the faithless modern world
    The heart fails.
No cure can be found
From a doctor or shaman.
The stores of ancient medicines are spent.
Leaving the world-wear of this age in decline.

The cure is not to be found in a capsule or ampoule;
The cure must instead come from *within*.
Synthetics are administered to the body
Yet they cannot imbue the heart
Or re-weave the fibers of the fraying mind.

And so,
We whose hearts are sickly
        Must seek.

Trapped for far too long
In this post-revelations civilization,
I have become my fears.
The fever of doubt ever-rages within me
And in fits of pessimism I become death.

I have become my scars.
Each word I speak
And each choice I make
Is a reaction to pain felt
Or a result of the worries that linger.

The instincts of my heart
Have been replaced by recoils.
I have become a knotted mass
Of insecurities, trauma and doubts.

The torturers that haunted my every step
    Were cast out long ago
Yet now I find I punish myself;
For my self-worth was taken from me
I am left to suffer the constant ridicule
Of the one who hates me
    Dwelling here—within me.

Afflicted I fall at your feet.
Lonely, ostracized, longing for connection,
Wanting, not only for another to love me
But to be able to love myself,
    I reach to you.

Unable to go on…
Dying from the inside out
    I call to, Mother…
    I call to you, Father.

It is I—the idealist—
    The believer,
After being eaten alive
By those with bitter tongues
And left unable to see the beauty
    In life or within myself.

It is I—the child—
    The vibrant soul
Who once bound into your arms each morning
And reached for your embrace each night,
Aged by the ware and strain suffered
In that heathen place where love is dead
      And you are disregarded.

Moving through the still expanse
That lies between time and space
I am carried across the ages
    —Across the distance between us—
By this heaven-sent vessel,
I have finally returned to the shores of home.

Sweet Mother, hold me.
I fall into your arms,
Unable to hold myself up.

Wrap me in a thick quilt
Sewn by thine own hand;
Then lay me down and stroke my hair.

I need not answers, Mother.
I need the restorative care
That only you can give.

After the horrors beheld
And the crimes partaken in
I need your wholesomeness.

After the ceaseless war that raged
Within and without
—I need your calm.

After the famine endured upon that wasteland
I need the meal fortified
With the goodness of thine own heart
To replenish me.

Your presence pulls me back from the madness.
Your softest touch has the power to break
The crushing grip the emptiness has had upon me.

Take me in, Mother.
Wash out my sore,
Infected wounds
With cool water.
Then bind me in soft cream colored linens;
Tight enough that I am held together
    And the hemorrhaging and unraveling
    Might fade.

Take off these cloths, Mother.
Divest me of these foreign garments
Woven by soulless steel spools
And drape around me long shawls of thickest cotton.

Pull over my limp head
A robe the color of the night sky
Woven from the wool
Sheared from our flock of black faced sheep
    That graze upon the hills
And let me reconnect with this—my past,
So that I can remember who I truly am.

Unlace these worn shoes, Mother.
Separate the soles of the boots
From the soles of my feet;
For the two have become melded with wear—
    Plastered together
    By the dried blood, sweat and sand
    That gathered as I traversed
    The scorched plains of the Hell.

Take up the shears and cut the hide laces
I used to keep these boots upon my feet
And in doing so let the time of journeying end…
Let there be peace and contentment;
For I know that this is where I shall settle—
    I know this is the place
    I always should have remained.

Gather water from the streams
That feed the lands of our home,
Pour it from the red clay pitchers
Over the split skin of my worn feet.

Then likewise bind them in soft linens
And lay me down into my bed,
Blow out the guttering candles
And let me take my rest.

Let me sink into the comfort
Of this place to which I belong.

Pull the blankets in around me, Mother.
Ease my head into the softest goose feather pillow.
Wipe the lingering tensions from my face
With a gentle pass of your loving hand
Across my forehead and down my cheek.

Then sit with me, Mother,
Until the images of the demons
Encountered in that place are forgotten.

Leave me in this cradle, Mother,
That I might reenter the womb of our home
And be restored—
    Encompassed by your care,
    Protected by your love
    And nurtured by the connection that runs
    From my heart to yours.

And finally,
When the new dawn is approaching,
Bid Father to come to me;
For he is the only one who can help me now.

Wake me, Father, to a new day.
Come remind me who I am.
Come guide me back unto myself.

I made it home, beloved Father.
I returned.
Yet I am still plagued
By the lingering sicknesses
Inherent to the wasteland I crossed.

I made it home, Father
Yet the tumors of bitterness,
    Guilt and pessimism
Must now be dissolved.

Help the weakened,
Starved parts of me
Regain strength.
Help the broken part of me
    Be set and refortified.

Come, sit at my bedside
And feed my soul teaspoons of hope,
        Belief, truth, and meaning.

Few hearts are such potent sources
Of these things as you.
I am in need of you, Father.
I cannot recover
From this weariness on my own.

While walking the plains of the wasteland,
The desolate horizon I stared into day after day
Began to wear on me.

The ghosts
    —The walking dead
    Who dwell there upon that barren land—
They terrorized me, they stalked me, they berated me.
And even though I seemed to have escaped them,
Some displaced embodiment of their acts haunt me still.

I made it home
But I did not stride across the threshold
    Of this—our home.

No, I dragged myself across
    —Falling over it—
Upon my deathbed.
Yet I know that here dwells
Those who can heal me.

Help me, Father;
For I am broke and bleeding
And I know not how to stop the reeling.

The acts that I chose to partake in
Loop unrelentingly in my mind.

Did the choice I made
To leave this land behind
Curse me forever?
Or is there still a hope
That I can return to the life I had before?

Dearest Father, take me back to who I was
And in doing so show me who I still can be.

Help me up, Father,
Let us go for a walk across these lands…
Guide me through this healing place
And help me find my way back
        To the life I left behind.

THE H⬦MELAND

When I lived in the wasteland,
   —Dwelling in the ruins
  Of the environmental holocaust—
I dreamt of tall grass fields
Growing upon an ageless land.

I saw myself
   Sweeping through them as a healthy child
Bursting with the energy of life,
The sound of full-hearted laughter
        Trailing in my wake.

And later, as an adult,
I saw myself walking through them
With a matured grace running my hands through
The thick blades and heavy tips of the wild wheat,
Savoring the radiant fullness of my own heart.

The vision of the fields followed me,
As did the feeling emanating
From the lands they covered—
   A place of intensity,
    Wrapped in a calm ease…
Where everything is bright, crisp, and ripe.

A sanctuary of abundance,
Where the very air we take in sooths
And keeps us sharp—
    Awake and aware.
…An ageless land where the ways past
Are the ways still kept.

I had no explanation for
These recollections of a vibrant place
That was so very different
From the dullness that surrounded me.

Images of days spent in fields
That I had never been to,
Were shrugged off
As recollections of some past life
Spent in a long dead age.
…Or perhaps some place I had journeyed to
               In my imagination alone.

For years I had nothing more than that
    —A vague explanation—
A guess at the unexplainable.
But at last the context has surfaced.

Like placing a name to a face,
I remembered where those fields grew.
I remembered a world that once flourished
Before the modern one was erected.

…I remembered my true self
And started the long journey
To search for a place
Where that old world still endured.

After years of sifting through
    Religion, history, and myth
Trying to find the lost and hidden truths….

After living exiled and impoverished
As a heretic against the modern religion
    Of progress and greed….

After hiking the long trek
From the city to the secluded shore….

After years of waiting for a guide
To lead me across the sea,

I have made it home.

We drew our boat up
Upon the sands of the other shore.
Crossing through a wood line of virgin trees
—The youngest of which had stood for centuries
And upon the other side of the grove
We emerged into the open fields
    Where in my visions I had taken flight.

As we walked through the lands
The past became the present.
With each step taken
The foreign became the familiar.
And we began to realize
That we had not *discovered* this other shore,
We had returned home to it.

Homecoming is, at first,
A reawakening to the life lived before…
A reintegration into the reality
We knew only as a dream.
Homecoming is a reclaiming of a life
That was preserved for us to reenter,
When our journeying came to an end
And we returned to our senses.

Wading into the waist high golden fields,
The beams of a gentle sun
Streaked through the cracks in the silver clouds
Dominating the cerulean sky above,
Embossing the rolling vista below.

A warm happiness was carried in the wind.
As we walked these lands we were
Surrounded by a loving presence.

No weights were carried across these hills—
The miles did not drain us;
Rather, with each step taken,
           We were refilled.

Here the wellspring of enthusiasm overflows,
      Washing away fears, drowning doubts,
And putting out the wildfires of hopelessness.

The lands are alive—
    Each part of the scene before me breathes;
    The patches of each different color
    Course with a dazzling, fluid vibrancy.

Treading barefoot across the rich black earth
I can feel that same spirit of life passing through me—
    Up through my feet, into my legs,
    Migrating upward to my chest, where it pools
    To refill the reservoir of strength, hope and possibility
    Drained during the famine suffered upon the wasteland.

Continuing ever-upward into my mind
Clearing all torments, all dullness—
    Washing clean the film of weariness from my eyes,
    Allowing all the shades of color
    To return to my black and white existence.

For a lifetime I yearned for the emerald horizon,
Where no buildings would besmirch the serene path before me.

Leaving the high grasses of the meadows,
I moved up the slope of the hill
Where the scope of this otherworld opened full to me
And breathlessly, I beheld the earth as it was
Before the onset of greed disfigured it.

There, beyond the crest of the hill,
        I saw the face of my god.

The swollen blue lochs were afire
With the white sunlight
Blazing across their surface.

A nesting of thatch roofs
Were settled cozily at the base of a far off mount;
Curls of smoke rose from chimneys;
The stark whitewash walls of each cottage
            Were clearly visible…
Painted the color of freshly risen heavy cream.

The thick mossy blanket of the valley's grasses
Crept up the sides of the steep black tors,
    Which stood like megaliths
    Framing the threshold
    Of a wall-less, roofless temple.

Tufts of tall grass spotted the smoothly rolling lands;
While banks of low growing bushes and brambles
    Skirted the base of the sloping hills.

Taking in a breath of the salty air,
    Fragrant with heather and hearth smoke
I moved down into the lowlands.

…I am home.

THE VOYAGE BACK

The wasteland seems so far away.
But, at the same time,
Still some part of it was near,
As a twinge of lingering ties tugged on me
When I remembered those kin whom still dwell there.

A vision of the family we once were
Expanded within my mind,
Obstructing the scene before me.
And as I descended the mount I could sense that,
  Though I would wish it to be,
The time of journeying was not yet over.

"You are throwing yourself back into the storm."
My guide said to me,
As the black bow of our cutter
Broke through the dense, engulfing fog.

"Yes, I know." I replied,
Fully aware of what I was moving towards...
Clutching my loosely woven cloth bag to my chest
As the shore of the wasteland loomed back into view—
    A ghostly corpse of a world
    Sprawled upon the flat horizon.

"Then why go back?"
    This guide asked in a heavy voice,
Curious as to what could compel me
To return to Hell, after all I endured to escape.

"Because I can hear the screams of those caught there
    Reverberating within me
    And I can bear them no more."

Leaving the *other shore*,
    I return to the wasteland
With two talismans to keep me ever-connected
To the kin I leave behind and the ways in which we live:
    An auburn lock of hair from my own innocent babe
    And a thin silver rod in eternal bloom
    —A cutting from the *crann bethadh*—
The tree that roots at the axis
Between the seen and the unseen worlds,
Sheltering and sustaining
We who dwell there upon the sanctuary isle.

I knew I must take these keepsakes,
Not just to ease my longing
For those I must live without for a time
But as a way of protecting my personal truths;
For they are more than objects
Through which I shall recall love.
They are objects through which I shall recall myself
And shall remain firmly grounded
        Within my true life.

On a journey to a place such as this,
They are most needed;
For in this world I return to,
One can lose themselves in an instant.
As if the very place emits a distorting force
That robs us of our proper perspective
And all the truths we once knew.

It was perilous to risk a return.
Nevertheless, after taking the boat home
Unto that isle of eternal youth
    —That isle of living ancestors,
    Known by a different name
    Among each different peoples—
I made the choice to return.

After waiting a lifetime for the boat to come
I am doing what I never thought I would—
    I am returning to the wasteland
I devoted a lifetime to escaping.

Making the voyage back
Out of a need to ensure,
        One final time,
That none of the kin I lost to this place
           Can be retrieved.

I bring you a bushel of apples
From the bountiful homeland,
Oh starving Cousin.

I bring you clippings of white heather
And flowering thistle sheared from the moors.

I bring you a handful of purple hazel nuts
Yielded by the sacred grove.

I bring you a bundle of dried rose hips
Plucked from the bush that grows near our garden gate.
Come, let me make you a cup of tea from them
And let us reminisce of the comforts of home,
That you might remember the joys that old life gave to you,
And you might be willing to return there with me.

I bring you a tin of ashes gathered
From the hearth that burns at the center of our home;
Contained herein are the remains of the fat oak log
We burned on the solstice
To ward off the darkness of the approaching winter.

Come, breathe in the lingering aroma of the smoke
And let it remind you of the fullness had in the simple life
    Of working the rich soil
    And nurturing the loved ones
    Bestowed unto us by fate.

Take my hand, wayward Cousin
And remember the smells of the wildflower meadows.
Let go this need for nuts and bolts,
Iron and steel, bills and coins
And let me take you home to the woodland,
Where you can be surrounded by life once more.

Before leaving on my voyage back here to you
I banked the home fire with three dried tufts of peat moss.
Then made my way to the yew tree
Spiraling on the green at the center of our village,
Where I tore away the hem of my cloak
And tied it to a low hanging branch;
In a prayer that you might be healed
From your need for this detrimental way of living.

Does nothing from our past life move you?
Can you not recall the herds and flocks
Whose mews we once found so endearing?

Can you not remember the colony of puffins
That crowd upon the thin black isle
Beyond our eastern shore?

Can you not recall the chill that passed over us as,
Sitting fireside, we heard the lament
Of the loon break the calm of the night?

Can you not recall the bleating cry of the newborn lamb
As it stagers forward into the world
Dazed by the light and noise?

Can you not recall the moan of the ink black seals
Beached on the edge of the low rocks,
Near the spray of the breaking swells?

Can you recall the frothy salt moving in swirls
Across the surface of the slate gray waters,
As the tide filled the hollowed alcoves?

Can you not remember the parties had
On the nights of the harvest festivals?

Adorned trees festive with satin ribbons
Streaming from their fingertips.

The chaplets donned of golden barley,
Beaming like a halo around each face.

Young girls bestowed circlets of coiled willow branch
And flowering apple blossoms;
While the boys were knighted with crowns of wound ivy
    And white berried mistletoe.

Beacon fires dotting the length of the strand—
    An isle in celebration.
The new bounty shared by all,
In feasts had within the vaulted dining hall.

Sweet mead—fragrant with ripen berries—
Passed round the long tables in earthen pitchers.
Heavy clay plates—spun from red mud
    Pulled up by hand from along the riverbed—
Laden with honey cakes, hardy breads,
    Roasted meat and ripened fruits.

Come, let me take you home just as I was.

Come, step inside this circle I have drawn
With the tip of this silver branch;
Within it you will be safe
From the demons of greed that possess you.
And perhaps for a moment
You might return to the innocent one
You were when we were children.

Step into this place removed
And be transported back, dear Cousin

Can you not smell the brine churning
Among the kelp fields rippling in the tide
Just off the shores of our home?

Can you not hear the murmurings
That ride the tails of the wind encircling the isle?

Can you not see the bonfires of *Beltaine*
Lit in the four corners of the all-encompassing sacred wood?

Can you not smell Mother's haddock stew
Simmering on the hearth?

Can you not hear her voice
Calling us home?

Come, let us return and dust off the trunk
                    Of your former life.
Reopen it,
And let me make you feel human again…
Allow me to unburden you of these
Constraining poly-synthetic clothes
    That are all form and no soul
And draw around you one of your old robes
Of hand-dyed spun wool.

Here, put on the sheepskin boots
That our father sewed for you,
                Take my arm
And let us rediscover the beauty
                Of our homeland.

GRASSROOTS

The grassroots of humankind's spirituality
—This ecological faith—
Was founded not by prophets
But by peasants…
By those who had nothing
And so discovered the worth
In the naturalness growing around them.

A homely faith that is without arrogance,
The sacred objects of which
Are the ubiquitously growing
    Wood, grass, and berries,
    The many-formed breasts and birds
    And the life-guarding elements
            Fire and water.

The presence of Mother and Father
    —The divine beings—
Was discovered by those who dwelt in hovels—
    A rustic folk, uneducated
    But with a keen eye and open heart.

Farmers not scholars;
Family men not clerics;
Paupers of low class
With dirty hands, calloused feet,
Yellowing teeth and matted hair.
…A rustic folk, though heart-hardy.

It was there—
    While tiling the fields,
    Raking the muck,
    Feeding the pigs
    And tending the fire
That the rumination of purity began.

It was while
The warm wind blew,
The wildflowers bloomed
And the clockwork of the stars moved in turn
That these hardened folk
Felt the presence of some greater power
And began a contemplation of the sacred
That would be passed down
To every generation to follow after them.

In and amongst the lush green of the wood
They could sense some magnanimous spirit at work.
In the return of the spring after the winter
They could sense this being's merciful nature.

They knew of their small place
Within the intimidating grand scope of the world
And they recognized the being whose
All-encompassing hands
Hold all things in balance.

Gleaning an understanding of the unfathomable
Through what their eyes could see and heart could feel
The simple folk became the wise.

Their respect for the wild realm
    And the force governing it
Verged on reverence and,
    Alight with a desire to know the being
    Whose aura they sense within all things,
        Spirituality was born.

Centuries since the discovery
Of the sacred within the meadows,
The strongest faith is still had
By those of meager means.

We can try to dig up
The grassroots from which we sprung
But they go too deeply into us to ever be eradicated.

Generations later,
We are quite removed from the hovel
Yet we cannot change the blood
Of the herdsman and farmers
That runs through our veins.

We strive to escape our poverty
Yet we who have suffered it
Can be understood by no other kind.

I can understand a poor mother
Better than I can a rich man.
I began to rise from my poverty
Into the ranks of middle class
Yet found myself unable to relate to one
Who has never known the struggle.

Leading me to realize that I belonged in the slums
With the desperate and the dying;
For, though they have not a shilling,
They know how difficult this life can be.
And consequently, they are the only ones
Who can appreciate what my life had been.

A fraternity of the damned and disowned,
Only we can know what each other has faced.
Made brothers and sisters through our shared trials,
Which the imagination of the sheltered mind cannot fathom,
We can look into the eyes of one we meet
And see if he or she is kindred to us.

Visiting that lowest level of survival,
    Wherein our last remnants of pride are stripped from us
    As we are forced to beg for scraps worthy of only a mutt,
Leaves a mark upon one—a look about the eyes—
Which others who have also been there can recognize.

Preferring the harsh honesty
And rough living of the peasants,
I would rather be given a crust of bread by a beggar
Than be bought a banquet by a benefactor;
For the beggar gives from his heart,
While the rich man has made his money
By living without one.

Unable to flush the blood
Of the pleasant from my veins
I shall honor what I am,
    Return to the cottage
And perhaps find in that modest hut,
The place where I belong.

I seek to return to my humble beginning
To commune with the sacred in its primal form.

To depart the civilized world
Where we live boxed up away from the earth
And return to the dirt floored hovel.
Never again to touch the steel knobs of a faucet
   —One for hot and one for cold—
But to draw my water
From the streams and stone-walled wells
    Cool water in the winter,
    Warm in the summer.

Kept too long in these polyester sheets,
I miss my bed of prairie grass
And my blanket of the summer sun.

I miss washing the dust from my feet upon returning
From a barefoot walk along the well-worn paths.
And sweeping through the thick growth—
    The tips of each pronged leaf
    Bejeweled with morning dew,
Which soaks my shirt and pant legs as I brush past.

Come, dance around the bonfire with me
    —Surrender yourself,
Indulge the instincts you have long ignored.

Come, I have tilled the soil
Now you follow behind me
And sow the seeds.

Come, resume your place—
Walk out into the torrential rains
And be re-baptized into the natural faith.

Do not deny your rustic soul
That yearns for the smell of the hearth
And the feel of cool dirt between your fingers.
Remove yourself from your modern prison,
Which suppresses every instinct of your native self
        And become the old ways…
Embrace the folk from who you descend.

We built this modern world
Thinking it would fulfill us,
Only to discover that this way of life
Brings emptiness, not ease.

Yet instead of dismantling this world we made
And returning to the old ways
We suffered ourselves to stay
And let the great machine keep churning.

We coined the modern adage,
"You can't return home."
Condemning ourselves to a way of life
Where joy is seldom found;
Closing a door
That would have always remained open to us.

…A door that still can be reopened,
If only we admit that we are a people of the earth
And what we need to be fulfilled
Lies within the simple ways we left behind.

THE L☉NG PATH
☉F UNDERSTANDING

"Re-examine all that you have been told...
dismiss that which insults your soul."
— Walt Whitman[1]

The Personal Significance of Oak Wise in my Spiritual Journey

I do not hold a single faith. My spirituality is a braided cord of phi-
losophies, ideals, and personal truths gathered along my journey.
Over the course of my fifteen year spiritual journey I have progressed
through studies of Catholicism, Judaism, Buddhism, Druidry, and the
shamanic traditions. *Oak Wise* represents my passage through Druidry
and Celtic shamanism.

During my journey through these two particular traditions I gath-
ered two significant affirmations: Firstly, from Druidry, I was given an
affirmation that the earth is indeed humanity's church; nature is sacred.

Secondly, from shamanism, I was given the affirmation that the indi-
vidual is directly connected to the divine and therefore needs no interface
in order to commune with what is sacred.

While I did not put down roots in these two traditions, I look back
on my discovery of these philosophies fondly. My passage through Dru-
idry and shamanism yielded the aforementioned affirmations, which al-
lowed me to solidify long-held beliefs I had come to years before on my
own.

Since my curiosity concerning the *great mystery* first emerged, I have
believed that humanity is meant to gather an understanding of the divine
from firsthand experience. While I hold a reverence for the wisdom of
the past, I do not believe that insights of previous generations should go

---

1 Walt Whitman, *Leaves of Grass: The Original 1855 Edition* (Dover Publications, 2007), p. 8.

unscrutinized or unprogressed. True wisdom cannot be set in stone; for it is ever-evolving. Blind acceptance of unprogressed insight leads to a morally and spiritually stunted population. Leaving us with faiths such as Catholicism, Islam, and Judaism, which struggle to find their place in a progressing world as they cling to outdated and often misinterpreted truths.

It is my belief that if we follow the heeding of our heart—follow the instincts of the soul—we will be led to the answers we are in need of. I believe that every soul is born connected to the divine and through that connection we can come to understand the deeper unseen workings underlining our life and the world around us.

Overcoming Stigmas

Many years ago now, I decided to diverge from the mainstream faiths. Like with all journeys there have been high points and low points during my path. The way of a unique thinker can often be a long and lonely one.

My personal decision to move beyond the well-known religions and explore the "pagan" faiths, drew a great deal of criticism. Nonetheless, I concluded I must follow my heart to whatever end, regardless of stigmas.

The firmness of this conclusion did not make the path any easier. Not fitting into a box often means having no established house of belonging and no common ground with those around us. There were times in my life, early on in my spiritual evolution, where I chose to *smooth out* my spiritual views while in public just to give myself a moment where I could relate to those around me. Of course, any bonding from such exchanges was an illusion, but what can I say: I needed a momentary break from being a misunderstood voice railing in the descent.

We may, at times, find ourselves isolated by our uniqueness but this alienation mustn't stop us from heeding the whispers of our heart unto new revelation. We must not allow for stagnancy of knowledge, and knowledge will only be progressed by those thinkers willing to leave behind the mob and think beyond the margins. Followers of established religions cling to the texts of wise men and women who went *outside* the margins of the established ways to evolve the truths of those who came before.

When I published *Oak Wise* in 2010, I did so out of a desire to share a glimpse of old wisdom I felt had relevance for the current generations. In my opinion, one of the central truths of shamanism is that we are all connected to the ancient being we seek to understand. While living this life, we are all making a spiritual journey as we attempt to connect and maintain our authentic self and know our modest part in all that is taking place in the unseen.

Pushing Past Fatigue of Spirit

I will not pretend that being an independent thinker in this day and age is an easy task but, looking back through history, acting outside the mainstream has always held danger. The isolation, misinterpretation, and condemnation can wear on the heart and soul, leaving a once-robust idealist broken and disinterested.

I fell victim to such a fatigue. When I was younger I delved often into the depths of the *greater mystery* but now I know how isolating it is out there on the fringes—so few people venturing out from the safe shoals of shallow thinking.

During the low points of my spiritual search I found myself robbed of my courage to wade out and explore. The unknown was once so seductive and my curiosity so deep, that I found myself driven to get up at dawn and ponder the meaning of this life as the sun rose. Now, I think, if I were to speak honestly, I would have to confess that I have lost my belief that the mysteries can ever be fully understood. This life—my purpose, my path, the meaning behind it all—I once felt that it was all *knowable* but now it often feels like I am a child playing with puzzle pieces that are never going to fit together.

I would give it all up—throw up my hands and declare this life aimless and senseless—if it weren't for the small things that occur, which point to something deeper at work. Such as connecting with rare like-minded souls among the masses of indifferent or experiencing one of those fleeting moments of fierce clarity, we all desire.

Thirty years into my life, I certainly have more questions than I did in the beginning. I have few answers and, I will admit, the holes in my understanding torment me. Yet as tired as I am of demanding answers from the Fates who seem all but deaf, I still find myself ever-aware of the small events that whisper of deeper-workings; even if what these whispers speak of is beyond my current comprehension.

In the end, all we can do is follow the progression of our faith throughout this life—wherever it may take us. And hope that, at some point, all the fragments of experience and insight will come together—if only for the fraction of a moment—to form a greater understanding.

—L.M. Browning
Connecticut | Autumn, 2012

GL⊙SSARY

Many of the poems use words of particular cultural and literary reference. There are also Celtic, mystical, and historical references and other terms, which may be unfamiliar to some readers.

Anam Cara — Gaelic. Rough Translation: Soul friend. *Anam* (soul) *Cara* (friend). A spiritual binding of two souls. A person with who one can share their innermost thoughts and feelings. The deepest relationship that can be had between two people; one that is not effected by distance, time, or death.

Arcadia — Mythological utopia. An idyllic place of unspoiled wilderness. Inhabited by an innocent people uncorrupted by greed, pride, or violence.

Bard — Gaelic. Bards are keepers of history and lore. Composers of poems, songs, and tales. In the Druid tradition, the training of a bard was known to take several decades and consisted of learning hundreds of poems, stories, and histories so to be able to recite them from memory. They trained with the intention of preserving insights and histories as well as to educate and aid those they may encounter.

Beltaine — (byel-tin-yuh) Gaelic. Festival in the Druid faith that celebrates the coming of summer. Marked by bonfires, feasts, dancing, and other rituals. During this festival the new years crops are blessed and the herds were once again driven out to the grazing grounds. The festival is held annually on May 1st.

Blood (Concept of soul residing therein). — It is an old Celtic belief that the soul may reside in the blood. Drinking the blood of a fallen comrade or kinsman, while beyond our modern comprehension, was perceived as a way of bonding two people.

Birch (Significance in Druidism). — One of the trees held sacred in Druidry, the birch represents purity and renewal. Purity in its white bark and renewal in that it sheds its bark each season.

Caim — (kime) Celtic. Rough Translation: Encircling or encompassing. An ancient ritual wherein a circle is drawn around a person who is in need of divine protection or guarding. The circle acts as a defensive boundary creating a sanctuary within for the one in need. The *caim* is performed by those haunted or hunted by evil, plagued by illness, or in other such danger.

Chaplet — Wreath or circle of flowers worn on the head.

Clootie Tree — Scottish. Rough Translation: A piece of cloth. A ritual wherein a strip of cloth is tied onto the branch of a tree. One performs this ritual out of a desire to heal themselves or another of an ailment. A wish is made with the rag then it is tied onto a tree branch, as the rag decays over time in the elements so too the disease decays. The Clootie trees are usually located at the site of a sacred well (Clootie Well) said to have healing waters or on the green at the center of the village.

Crann Bethadh — Gaelic. Rough Translation: Tree of life within Celtic tradition. The tree of life is an icon/concept that is present within almost every religion and culture in the world. The myth holds that the tree of life is rooted at the center of the otherworld, in some cultures it is known as the "axis mundi" (the center of the world/world pillar). The tree crosses the three plains: underworld (roots), surface/earth (trunk) and upperworld/sky (crown). The tree symbolizes many things in Druidism: abundance, rebirth, immortality, dignity, and shelter. They are also known as doorways to the otherworld. Mortals have been known to fall asleep under trees and wake in the otherworld.

Daemons — Guardian spirit. A spirit supposed to look after a person or place.

Druid — Celtic. Rough Translation: Oak knower or oak wise. A Druid was a class of learned practitioner/priest in the Celtic/Gaul community who focused upon the establishing and keeping of laws as well as overseeing religious ceremony. Druids were one class of practitioner in Celtic antiquity, others being: Bards, *Filid* (visionary poets), and Vates (seers/healers). This definition "Celtic" loosely meaning the peoples of Ireland, Scotland, Britannia, Wales, and Gaul. In modern day context the words Druid/Druidism have become a broad term encompassing all those who follow the earth-based faith of the Celtic people.

Drumming — Drums are prominent in almost ever aboriginal culture and faith. In the shamanic faith drumming is believed to induce an altered state of consciousness.

Earth-based Faith — A "religion" or philosophy with beliefs that revolve around the earth/nature or promote connection to the divine through the earth/nature. In an earth-based faith the earth/nature is held with the same reverence as the divine, whether in the form of worshiping the spirit of the earth itself in the form of a mother goddess or holding the earth to be in and of itself, sacred. Some examples of earth-based faiths are: Druidry, shamanism, the aboriginal faiths, the animist faiths, and the Native America faiths.

Evergreen (Significance in Druidism). — The evergreen is a symbol of rebirth but also of life's endurance. When all the other trees lose their leaves and appear dead throughout the winter months the evergreen does not. It was/is a Druid/pagan tradition to bring in cuttings of those bushes and trees that remain green year-round (i.e.: holly, pine, evergreens) and display them in the house throughout the winter months as a reminder that spring shall return.

Faery Folk — Beings of the otherworld who venture into our own realm sometimes to enchant us or at times to lead us back to the otherworld.

Fire in the Head — A state of increased/acute awareness, idealism, inspiration and the desire to act, often associated with shamanism.

Gaea — also spelled "Gaia" or "Gea". Greek. Rough Translation: Mother Earth / *Ge* (earth or land) *Aia* (Mother or Grandmother). Greek goddess personifying mother earth. Has also come to represent the feminine divine/the mother goddess.

Hazel (Significance in Druidism). — A hazel tree or grove thereof are said to grow at the heart of the otherworld. The nuts that drop from the hazel tree are meant to be gathered up and eaten; they represent the truths/realizations for which we all long.

Hiraeth — Welsh. Rough Translation: Homesickness. The longing to return to the homeland. The longing for something once had now lost; whether had in a previous life and now lost or had in the present and left behind or destroyed.

Inner-Directed — One who is governed by personal beliefs. One who is guided by personal beliefs rather than by norms imposed by society.

Megaliths — Large standing stone(s). An enormous stone, usually standing upright or forming part of a prehistoric structure.

Mistletoe — Mistletoe has been held sacred by many cultures including the Celts and Norse who believed the plant guards against evil. It is also held as a symbol of peace.

Moor — Old English. Wild area of countryside. A large uncultivated treeless stretch of land covered with bracken, heather, coarse grasses and/or moss.

Mysticism — The belief that knowledge of the divine, reality, and other truths are achieved through personal communication or union with the divine rather than through the interface of a formal religion.

Otherworld — The otherworld has been viewed many different ways through the eyes of each different culture. The vision focused upon in this book is that of the hidden isle or a land that is set apart from time, where a wise people live in balance with the earth. The concept of this *sacred isle* is present in many cultures throughout antiquity and has been given many different names in each tongue. In the instance of Celtic mythology the idea of the hidden isle crops up repeatedly in the form of: *Tir na nÓg, Mag Mell, Tir Tairngire,* and *Avalon.*

Pyre — A pile of burning material, especially a pile of wood, on which a dead body is ceremonially cremated at a funeral.

Rowan (Significance in Druidism) — Rooted in the Norse word: *Raun.* Rough Translation: To get red. And the Gaelic word: *Rudha-an.* Rough Translation: Red one. Also known as mountain ash. The rowan tree is believed to have protective properties, in that it will protect one from malicious presences/spirits. It is held in Druidry to be sacred and in Celtic myth is said to have been transplanted from the otherworld into this by a people called *Tuatha Dé Danann.*

Shaman — There are many variations of shamanism through the cultures of the world but all shamans are usually healers, mediators, and seers. Nearly all cultures have their own shamanic faith. Shamans themselves act as conduits to the unseen and aid their village through insights. One does not usually choose to become a shaman; rather, one is chosen. Shamans commune with the ancestors (those now in the unseen plain/otherworld). Shamans are attuned souls with the ability to *walk between the worlds.*

Thin Place — A place, time of day, or specific day within the course of the year where the veil between this world and the otherworld is *thin*. Making it easier to commune with those in the otherworld or even for the boundaries to be crossed and journeys to be made. Examples of thin places: Ancient ruins, crossroads, shorelines, within the mists/fog, or at a river crossing. Examples of thin times of day: Sunrise and sunset. Examples of days throughout the year where the veil is thin: *Samhain* (All Hallows Eve) and the other Druid festival *Beltaine.*

White Buffalo — The birth of a white buffalo is considered an auspicious sign among the North American tribes. Regarded as sacred.

Yew (Significance in Druidism) — Native to Britain. A species of evergreen, the yew tree has long been viewed as a symbol of resurrection and immorality. The yew tree can live for upwards of four thousand years. Unlike other trees, the lower branches of a yew grow down and become new shoots that are, over time, folded into the trees ever-fattening trunk. Yew trees are known for their healing abilities. The sick within the village would be taken unto a yew tree where they would tie a strip of cloth upon one of the branches (a Clootie) making a wish for good health. Or in other instances the ailing person would be walked around the base of the tree and the healing properties of the tree would penetrate them.

Yule Log — The Yule-Tide season originally comes from the pagan/Druid winter festival; later it was folded into Catholicism after the Isles of Britannia and Ireland were converted. The Yule log itself is a fat oak log burned in the hearth on the winter solstice to ward off the approaching coldness. The tree from which the log is taken is chosen with tremendous care. It is known in Ireland as the *Bloc na Nollaig* or "the festival block". The fire burned bright during the season of darkness and the ashes of the burned Yule log would be kept in a tin or box as a reminder that warmth shall return—spring shall return.

L.M. BR⊙WNING

L.M. Browning grew up in a small fishing village in Connecticut. A long-time student of religion, nature and philosophy these themes permeate her work. Browning is a award-winning author and wildlife artist. In 2010 she wrote a Pushcart Prize nominated contemplative poetry series: *Oak Wise, Ruminations at Twilight* and *The Barren Plain*. In late 2011 she celebrated the release of her first full-length novel: *The Nameless Man*. She is a graduate from the University of London and a Fellow with the League of Conservationist Writers. In 2010 she accepted a partnership at Hiraeth Press—an independent publisher of ecological titles. She is Co-Founder and Associate Editor of *Written River: A Journal of Eco-Poetics* as well as Founder and Executive Editor of *The Wayfarer: A Journal of Contemplative Literature*. In 2011 Browning opened Homebound Publications—an independent publisher of contemplative literature based in New England. For more information visit: www.lmbrowning.com

FOREWORD BY ALAN COOKE

A native son of Ireland, Alan Cooke is an Emmy-winning actor, writer, and filmmaker. In 2002 Cooke made his New York stage debut in a show about the events of 9/11. After spending five years living in America, Cooke made a multi-award winning documentary about his journey as an immigrant in New York called *Home* featuring leading actors such as Liam Neeson, Mike Myers, Alfred Molina, Susan Sarandon, Rosie Perez, and the acclaimed Irish novelist Frank McCourt. *Home* debuted to immense praise and went on to win several awards: Winner—Best Documentary Magners Boston Irish Film Festival, Official Selection Galway Film Fleadh, Official Selection San Francisco Irish Film Festival, and Official Selection Chicago Irish Film Festival. Finally, in 2009 *Home* was nominated for three Emmy's including Best Documentary. Cooke went on to win an Emmy for Best Writing in a Documentary.

Cooke is the author of two works of non-fiction: *Naked in New York,* which chronicles his time in New York City and *The Spirit of Ireland*—a memoir following his time in his homeland. Cooke hopes to turning *The Spirit of Ireland* into a documentary exploring all that is beautiful, sad, powerful, poetic, and heroic about Ireland and its people. Returning to his native country, Alan wanders throughout the landscapes of Ireland seeking to capture the Irish identity.

HOMEBOUND
PUBLICATIONS

GOING BACK TO GO FORWARD is the philosophy of Homebound. We recognize the importance of going home to gather from the stores of old wisdom to help nourish our lives in this modern era. We choose to lend voice to those individuals who endeavor to translate the old truths into new context. Our titles introduce insights concerning mankind's present internal, social and ecological dilemmas.

It is the intention of those at Homebound to revive contemplative storytelling. We publish introspective full-length novels, parables, essay collections, epic verse, short story collections, journals and travel writing. In our fiction titles our intention is to introduce a new mythology that will directly aid mankind in the trials we face at present.

The stories humanity lives by give both context and perspective to our lives. Some older stories, while well-known to the generations, no longer resonate with the heart of the modern man nor do they address the present situation we face individually and as a global village. Homebound chooses titles that balance a reverence for the old wisdom; while at the same time presenting new perspectives by which to live.

CPSIA information can be obtained at www.ICGtesting.com
Printed in the USA
BVOW071556241012

303811BV00003B/1/P